BLACK INVENTORS SCIENCE WORKBOOK

AUTHOR'S ACKNOWLEDGEMENTS

I would like to sincerely thank my family for their support throughout the journey of bringing this project to fruition. To my wife - Senemeht Amen-Olatunji, thank you for your unwavering support and dedication to the goal of publishing my first book. To my children - Akhita Maaungkh and Sheta Raab, thank you for being an inspiration and the first students to read and critique the book.

To Keith Holmes, you served as the spark for the idea of a Black Inventors Children's Workbook. Thank you for not only initiating the idea but serving as a guide and mentor. Your depth and wealth of knowledge into the world of inventors of African descent from around the world has assisted me in navigating the world of writing and publishing.

I also thank HOBAE Designs & Illustrations for bringing your amazing design talents to this project. You saw the vision of my early concepts and brought them to life.

PUBLISHING

Published by Level Genius Media, LLC
Copyright © 2019 by Level Genius Media, LLC
All rights reserved. No part of this publication may be reproduced or transmitted in any form or by any means, electronic or mechanical including photocopying, recording or any information storage and retrieval systems, without written permission from the publisher.

ISBN #: 978-0-578-60176-2

Front Cover Design: HOBAEDesigns
Book Illustrations: HOBAE Designs

Level Genius Media, LLC
Brooklyn, NY 11225

ENERGY & SPEED

Chapter 1

- Vocabulary
- Themes and Objectives
- Introduction to Lonnie G. Johnson
- Getting to the Core
    - Energy
    - Moving Matter
    - Kinetic Energy principles
    - Potential energy principles
- Exercise: Kinetic Potential Energy Exercise
- The Science and The Invention: Lonnie G. Johnson and The Super Soaker
- Investigation and Exploration: Experiment Using the Principles of Energy and Speed
- Writing in Science: Writing Exercise using the Principles of Energy and Speed
- Chapter Review: Energy and Speed

ENERGY TRANSFER

Chapter 2

- Vocabulary, Themes and Objectives
- Introduction to Dr. Patricia Bath
- Getting to the Core
    - Energy Transfer
    - Wave Energy
    - Sound Energy
    - Radiant Energy
    - Thermal Energy
- Energy Exercise
- Writing in Science: Energy transfer
- The Science and the Invention: Dr. Patricia Bath and Laser phaco Probe
- Investigation and Exploration: Experiment Using the Principles of Transfer Energy
- Chapter Review: Transfer Energy

COLLIDING OBJECTS

Chapter 3

- Vocabulary, Themes and Objectives
- Introduction to Andrew Jackson Beard
- Getting to the Core
    - Energy and Collision
    - Momentum
    - Collision
- Writing in Science: Using the Principles of Colliding Energy
- The Science and The Invention: Andrew Jackson & the Coupling Device
- Investigation and Exploration: Experiment Using the Principles of Colliding Energy
- Chapter Review: Transfer Energy

ENERGY CONSERVATION

Chapter 4

- Introduction to William Kamkwamba
- Getting to the Core
    - Conservation of Energy
    - Writing in Science: Writing Exercise Using the law of Energy Conservation
- The Science and The Invention: William Kamkwamba and Windmills
- Investigation and Exploration: Experiment Using the law of Energy Conservation
- Chapter Review: Energy Conservation

Welcome!

Philosophy & Rationale For the Curriculum

The curriculum has been designed for students in grades 4 and will address Next Generation Science Standards and ELA Common Core State Standards at this grade level. The rationale for the curriculum is based on numerous studies that indicate teaching children of African descent about their culture while involving them in activities that fosters racial pride and social connection can help to offset the discrimination and racial prejudices they are likely to face in the outside world. Additionally, African- American children who have a proud, informed and sober view of race are more likely to attain academic success (Wang & Huguley, 2012).

Objective

This workbook aims to develop students' knowledge of key contributions made by inventors and scientists of African descent in various areas of technology and science. Students will also have the opportunity to practice literacy skills by working through the informational texts, answering text dependent questions, conducting research and producing writing in various formats. The cross-content design will allow students to reinforce learning of grade level science standards in the areas of general life, earth and physical science as well as engineering design. While progressing through the workbook set students will also be exposed to the latest developments and advancements in STEM based on the science principles they have learned about. The ultimate goal of this experience is two-fold:
(1) To ensure students have a deep foundation and understanding of the science skills and concepts for their grade level

(2) to inspire students to consider becoming scientists themselves through the stories and lives of inventors who are or have used these science principles to change our world. And by exposing students to the innovations and breakthroughs that are changing our world now.

Chapter 1

ENERGY
Speed & Energy

Chapter Vocabulary
- Energy
- Speed
- Force
- Potential energy
- Kinetic Energy
- Mechanical Energy

Section Theme
Use evidence to construct an explanation relating the speed of an object to the energy of that object. 4-PS3-1

Objectives
- Define and use scientific terms related to energy and speed
- Apply scientific concepts learned
- Explain the significance of the scientist's invention/ innovation
- Demonstrate an understanding of the scientific concepts behind the invention

LONNIE G. JOHNSON

Lonnie G. Johnson is a scientist and inventor who has worked for the U.S. Air Force and the NASA Space Program. Johnson was born in Alabama in 1949.

Growing up, Johnson was always building and experimenting with science, including the time he created a rocket fuel that exploded and burned part of the kitchen in his home. Lonnie continued to display his inclination for science by entering and winning first place in a national science competition.

In his later years, Lonnie Johnson went on to earn a master's degree in nuclear engineering from Tuskegee University. Mr. Johnson is best known for his invention the Super Soaker, a powerful water play gun that changed the water toy industry. Lonnie Johnson invented the Super Soaker while working on a design for a heat pump that would use water instead of Freon, a chemical that is harmful to the environment. During testing of the heat pump, Johnson caused a powerful stream of water to shoot out of a piece of tube with a nozzle and the idea for the Super Soaker was born.

10 years after its release there have been over 200 million Super Soakers sold. This invention remains as one of the 20 top selling toys of all times. The Super Soaker was not Johnson's only successes—he currently holds over 80 patents and was inducted into the Inventor Hall of Fame. Lonnie Johnson is presently working on an invention that will improve the use of solar energy for electricity.

Chapter 1

ENERGY
How Speed Effects Energy

Core Ideas
- The faster a given object is moving the more energy it possesses.
- Energy can be transferred in various ways and between objects

Lonnie G Johnson
was able to create a water toy that could shoot water further than any other water gun before it. How did he use the science of speed and energy to create the Super Soaker?

ENERGY
When you hear the word energy what do you think about? Do you think about the energy it takes to move a car or power an electronic device? Or the energy a pitcher uses to throw a baseball? All of these are different types of energy. Energy is the ability to do work (create a force) that will cause change in matter. The more energy something has the greater its ability is to do work or to make things happen. One way we can measure something's energy is by looking at its speed, or how fast it is moving. The faster a given object is moving; the more energy it possesses.

Example: A bowling ball will move faster if the bowler uses more energy to throw the ball, and it will move slower if the bowler uses less energy. The faster the ball is moving the more energy it has, which means it can do more work. In this case an energetic throw knock down the bowling pins.

Explain how the roller coaster's speed gives it the energy to move along the tracks. What would happen if the roller coaster had less speed?

_____
_____
_____
_____
_____
_____
_____
_____
_____
_____

Chapter 1

ENERGY
Core Ideas

KINETIC ENERGY

When you hear the word energy what do you think? Energy is important because it is what causes everything to move. We can find energy in many forms. Although we say that we use energy, it is actually being transferred from one form to another through a force. Force is a push or pull upon an object that changes the motion of the object. In order for Johnson's Super Soaker invention to be a success he relied on the energy of motion. This type of energy is called kinetic energy. Kinetic energy is the energy an object has because of its motion.

This gazelle is using **Kinetic Energy** as it jumps and moves through air

POTENTIAL ENERGY

Another type of energy used in the Super Soaker is called potential energy. Potential energy is the energy an object has because of its position or state. Potential energy is stored energy. For example, if you hold a ball above your head the ball now has potential energy. The energy stored in the ball is the result of its position of being off the ground and over your head. If you drop it or throw it, the ball now has kinetic energy as it moves toward the ground.

With the ball above his head the man in this picture just created Potential Energy. The ball can be dropped or thrown to create **kinetic energy**

Draw 3 pictures showing the steps of something going from potential energy to kinetic energy. Write a caption to explain each picture.		

Chapter 1

ENERGY
The Invention & the Science

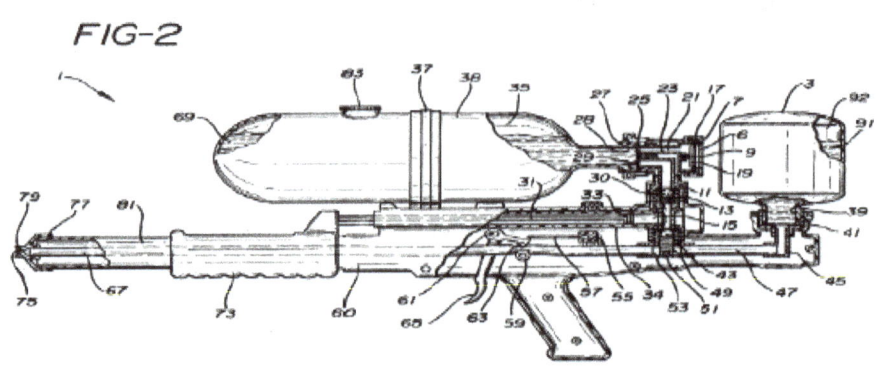

Our core idea tells us that the faster a given object is moving; the more energy it possess.

- The Super Soaker works by a pump system that transfers the kinetic energy of using the hand pump-to-pump air into a chamber.

- Once the air is pumped into the chamber, the kinetic energy of pumping is transferred to potential energy. (*Remember potential energy is stored energy because of an object's position*.) In the water chamber the air is pushing against the walls of the chamber as well as the water in the chamber creating pressure.

- The more air that is pumped into the chamber the greater the potential energy that is stored because the air cannot escape and pushes harder against the water and the walls of the water chamber.

- When the water is released the potential energy now becomes kinetic energy and the water shoots out with great speed, travelling longer distances.

- The scientist Lonnie Johnson was able to create more speed as the water comes out the water gun which gave it more energy.

Investigation and Exploration Using the Principles of: Speed and Energy

Procedure

- 1. Take a cardboard tube and cut it straight along one of its (long) sides. Then cut it along the other side so that you end up with two long pieces that are each semicircles. You will use one of these pieces as the ramp for the marble.

- 2. Place the bottom section of a milk carton at the bottom of the ramp to catch the marble

- 3. Use a marker to mark a starting line across the high end of the ramp, about one-half inch from the end. Procedure:

- 4. Starting at the line, release the marble down the ramp.

- 5. Measure the distance that the marble moves the carton. Record the distance.

- 6. Change the marble or the height of the ramp to see how the distance changes.

What you will need:
3 marbles of different sizes and/weight. A cardboard tube, a ruler, milk carton and identical sized books for the incline.

Questions:
-How far did each marble travel?
-How did the height of the ramp change the potentialenergy of the marble
-which marble had the most kinetic energy

Chapter 1

ENERGY
Writing in Science

Imagine you have gone on a trip to the circus. Write an explanation of one of the performances that uses the principles of energy and speed. Additional research may be done to complete the answer.

- Describe how the performance used energy and speed. Be sure to use the words kinetic energy, potential energy, force and speed in your description.

- Explain what would happen if there was a change in speed

Writing Exercise

Did you Know?
Shanghai Maglev Train
China is home to some of the world's fastest trains. It should come as no surprise that China is where you can find the fastest train in the world—The Shanghai Maglev Train. The train does not ride on rails like traditional trains. Instead it levitates (floats) above the tracks using magnetic energy. The train runs on electricity and can go as fast as 430 mph, but on average it travels at speeds around 230 mph.

Chapter 1

ENERGY
Chapter Review

1. When something has the ability to do work or cause change it has?
a. speed
b. energy
c. power

2. What is speed?
a. how far an object travels
b. how much time it takes an object to travel
c. how fast an object moves
d. the energy something has because of its position

3. A baseball player pulls his bat back getting ready to swing at the ball. What kind of energy is this?
a. electromagnetic energy
b. kinetic energy c. speed energy
d. potential energy

4. The energy an object has because of motion is called?
a. kinetic energy
b. gravitational energy
c. potential energy
d. speed energy

5. What has the most kinetic energy?
a. a boy holding a ball
b. a ball thrown across the field
c. a ball slowly rolling across the field
d. a ball on the ground

6. The push or pull on an object is:
a. a force
b. momentum
c. movement
d. speed

Read the questions carefully and provide the best response using complete sentences

7. Identify two experiences that helped Lonnie Johnson become a scientist and inventor? Why do you think these events were important to him becoming a scientist?

_____
_____
_____
_____
_____
_____
_____

8. Use the science of kinetic and potential energy to explain how Lonnie Johnson invented a water gun that could shoot water over a longer distance.

_____
_____
_____
_____
_____
_____
_____
_____
_____
_____

Chapter 2

ENERGY TRANSFER
How is energy moved from one place to another?

Chapter Vocabulary
- Energy
- Energy transfer
- Sound
- Light
- Heat
- Radiant energy

Section Theme
Make observations to provide evidence that energy can be transferred from place to place by sound, light, heat, and electric currents. 4-PS3-2

Objectives
- Describe wave energy, thermal energy, and electrical energy
- Explain how energy is transferred
- Explain the significance of the scientist's invention/ innovation
- Demonstrate an understanding of the scientific concepts behind the invention concepts behind the invention

DR. PATRICIA BATH

Dr. Patricia Bath is one of the greatest scientists of our time. Dr. Bath was born in 1942 and raised in Harlem, New York. Dr. Bath attended school in New York City and graduated from high school in two and a half years. In 1959 Dr. Bath attended a summer program for biomedical science. Shortly thereafter she began her college education at Hunter College in New York where she earned a degree in chemistry.

After that Dr. Bath attended Howard University Medical School where she earned a medical degree with honors. Dr. Bath went on to begin a great career in ophthalmology. Ophthalmology (op - thul - mal - o - gee) is the science that studies the parts of the eye and diseases of the eye. Dr. Bath contributed much to the medical field working for many universities and institutions. While working as an eye doctor she became interested in a disease called cataracts. Cataracts is a disease that causes cloudiness in the lens of the eye that makes vision blurry and can cause blindness.

Her interests in cataracts developed when she had the opportunity to work as an intern at Harlem Hospital while also attending a fellowship program at Columbia University. She realized that many of the patients at Harlem Hospital, which were mostly black suffered from blindness, while fewer patients at Columbia University suffered from blindness. Her research discovered that blacks were more likely to suffer from blindness due to cataracts than others because of lack of access to adequate treatments.

In 1988 Dr. Bath invented an instrument called the "Laserphaco Probe". This instrument uses a laser to remove the damaged lens from the eye and was a safer and less painful method to treat cataracts. Dr. Bath became the first African-American female doctor to be given a patent for a medical instrument. Dr. Bath finished her career in medicine at UCLA Medical Center where she became the first woman Director of a postgraduate program in the United States.

Chapter 2
ENERGY TRANSFER
Getting to the Core

Dr. Bath invented the Laserphaco Probe, a device that helped countless people avoid blindness. How did Dr. Bath use the science of energy transfer to design her life changing invention?

• Energy can be moved from place to place by moving objects or through sound, light, heat or electric currents.

• Energy can be transferred in various ways and between objects.

ENERGY TRANSFER
We learned last chapter that the amount of speed an object has lets us know how much energy it has. When we see a car driving down the street we know it has energy because of its speed. There are other types of energy that we experience everyday as well. Look around you right now—do you see lights, hear sounds, or feel any heat? Whenever there is motion, sound, light, electricity or heat—there is energy. The energy in light, sound, heat, and electricity is also caused by motion.

WAVE ENERGY
Wave energy is called wave energy because it travels in waves. Most of the time we do not see wave energy except for certain types of light energy. Sound and light are two types of wave energy. Actually light energy is a type of wave energy called radiant energy. There are other forms of radiant energy that we cannot see such as X—rays, ultraviolet rays, infrared, and others. There are also sound waves that we cannot hear.

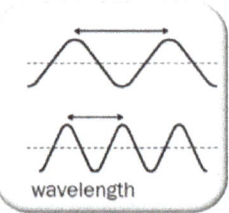
wavelength

What makes each sound different or so that we cannot hear it at all? What makes a light visible or invisible? There are two parts of a what that change the type of sound or radiant energy it produces. The first is the wavelength. Wavelength is the distance between two peaks of the wave. The second aspect that changes a wave is the frequency. Frequency tells us how many waves pass a certain point in a certain amount of time, for example in a second or minute.

SOUND ENERGY
Sound energy travels through sound waves. Sound is created when something moves back and forth causing a vibration. The vibration moves the air molecules creating sound waves in the air. We can hear the sound only if the sound waves can reach our eardrums which then causes the eardrums to vibrate. Energy is transferred from the vibrating object to air molecules and then to the eardrums. This is when a person hears the sound.

Did You Know?
Topaz Solar Farms In 2014 one of the largest solar farms was built in California. The Topaz Solar Farms powers almost 160,000 homes in the area. Solar farms are collections of solar panels built on large acres of land that power local homes and businesses. California is working to get 50% of its energy from renewable energy by 2030. That's a lot of energy transfer!

Chapter 2
ENERGY TRANSFER
Getting to the Core

RADIANT ENERGY

Light is another type of wave energy called radiant energy. Radiant energy includes visible light but also includes x-rays, gamma rays, infrared rays, microwaves, and ultraviolet rays. These are all different types of electromagnetic waves. As we learned earlier if the wavelength and frequency of a wave change, then so does the type of energy it produces. Look at the chart. As you can see we use the transfer of light energy all the time— when we warm up our food in a microwave, or use a remote control, and we feel the infrared energy from the sun as heat. These are all ways in which energy is moved using light.

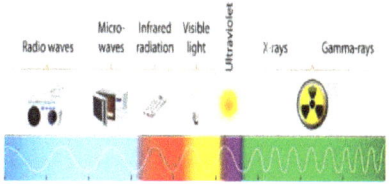

THERMAL ENERGY

The energy of heat is called thermal energy. Now that you are becoming an expert on the science of energy, you know that energy is produced by motion. If we go back to chapter one we learned that kinetic energy is the energy of motion. Everything is made of matter, and matter is made of molecules. All molecules have kinetic energy. The faster the molecules of a material move the more kinetic energy the molecules have and the hotter they become. The slower the molecules move, the less kinetic energy the molecules move and the cooler the material becomes. Heat can be generated by mechanical or physical energy, chemical energy, electrical energy, and radiant energy. Energy is the ability to do work or cause a change in matter. So when thermal energy is transferred it does the work of warming our homes, cooking our food, and keeping our planet warm.

ELECTRICAL ENERGY

By now we know that electrical energy must involve some form of movement because all energy comes from movement. Matter is made up of atoms. In the atoms are small particles called protons and electrons. Of course we never see atoms because they are so small (unless you use a special microscope). When electrons are moving from one place to another the movement creates an electrical charge. We use the continuous movement of electrical energy to create an electric current. The energy generated by the movement of electrons from one atom to another creates an electrical charge. The continuous movement of the electrical charge creates an electrical current which is then transferred to the things we use every day like lights, refrigerators, TV's and cell phones.

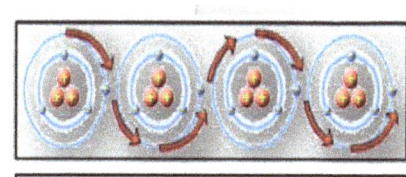

Chapter 2
ENERGY TRANSFER
Writing in Science

If you are in the room with someone who is playing the drums very loudly, how does covering your ears prevent the transfer of sound?

_____
_____
_____
_____
_____
_____
_____
_____
_____
_____
_____
_____

Directions: Heat or thermal energy can be created in many ways. Fill in the blank to identify how the heat is being generated in each situation.

- Physical energy
- Chemical Energy
- Radiant Energy
- Electrical Energy.

1. Rubbing your hands together makes them warm _____

2. Feeling the sun's heat on your face on a sunny day _____

3. Using a blow dyer to dry your hair _____

4. Burning natural gas to warm a home _____

Today many people are beginning to use solar energy to power homes, cars, and street lights. Write an informative piece explaining how the sun's energy is generated and how that energy is moved from the sun to power devices on earth. Include science vocabulary such as radiant, chemical and electrical energy. Additional research may be done to complete the answer.

_____
_____
_____
_____
_____
_____
_____
_____
_____
_____
_____
_____
_____
_____
_____

Chapter 2

ENERGY TRANSFER
Science and the Invention

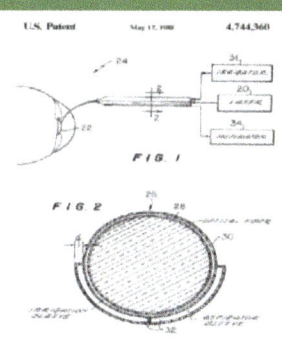
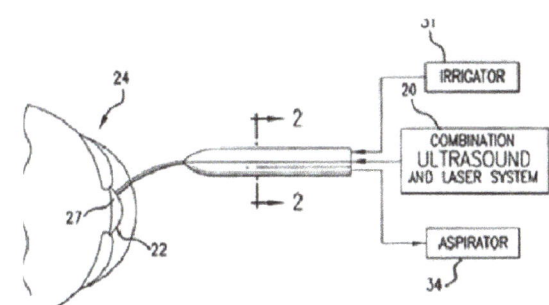

Our core idea tells us that: Energy can be moved from place to place through sound, light, heat or electric currents.

• Cataracts is a disease of the eye where the lens behind the iris and the pupil becomes cloudy making it difficult to see.

• Dr. Bath's invention - the Laserphco Probe, uses a laser that is inserted into the lens through a small incision (cut) to the eye.

• Using the science of energy transfer, Dr. Bath knew that the light energy of the laser could be transferred from the probe containing the laser to the clouded lens. The transferred energy would then destroy the damaged lens.

• Dr. Bath worked many years to find a laser that produced the right wavelength and frequency that would allow the damaged lens to be destroyed by the laser safely and then sucked out through a tube.

• Finally a new lens is inserted to replace the one that was removed.

Investigation and Exploration Using the Principles of: Energy Transfer.

Experiment: Colored Paper Transfer of Energy
Materials: 6 different colored construction paper (white, black, green, red, violet, yellow), 6 ice cubes, heat lamp or sunny area.

Preparation:
• Cut the construction paper into smaller squares (3 in. x 3 in.)
• Have student (s) predict which paper will cause the ice to melt the quickest.

Procedure:
• Place the construction paper squares under the heat lamp or in an area with equal sun exposure
• Place an ice cube on each surface
• See which ice cube melts first, see which melts last. Questions:
• Which surface caused the ice cube to melt first? Which ice cube melted last?
• Using what you learned about energy transfer explain what caused the ice cubes to melt at differentrates?
• How does this experiment relate to Dr. Patricia Bath's invention?

Chapter 2
ENERGY TRANSFER
Chapter Review

1. How is energy transferred from one place to another?
a. by waves
b. by moving objects
c. by sound, light, heat and electricity
d. all of the above

2. Why does a hot pizza get cooler when it is taken out of the oven?
a. The air makes the pizza get colder
b. Thermal energy from the pizza is transferred to the surrounding air
c. Thermal energy from the air is transferred to the pizza

3. What is the definition of thermal energy?
a. energy that comes from electricity
b. light energy
c. sound waves
d. energy that comes from heat

4. An electrical charge flows through?
a. waves
b. a liquid
c. current
d. the air

5. What is not true about wave energy?
a. wavelength changes the type of energy produced
b. energy is transferred through waves
c. wave frequency does not affect the energy produced
d. wave energy can travel through the air

7. What are cataracts, and how did Dr. Patricia Bath use radiant energy in her invention to treat cataracts?

_____

8. It took many years of research to find a laser that would safely destroy the damaged lens so that it could be replaced. What might happen when the laser hit the lens if it were too strong or too weak?

_____

Chapter 3

COLLIDING OBJECTS
How is energy transferred when objects collide?

Section Vocabulary
- Collide / Collision
- Energy Transfer
- Momentum
- Speed
- Force
- Potential energy
- Kinetic Energy

Section Theme
Ask questions and predict outcomes about the change in energy that occurs when objects collide 4-PS3-3

Objectives
- Explain how moving objects transfer energy when objects collide
- Apply knowledge of speed and energy to colliding objects
- Explain the significance of the scientist's invention/ innovation
- Demonstrate an understanding of the scientific concepts behind the invention

ANDREW JACKSON BEARD

Andrew Jackson Beard was an American inventor who is most famously known for his Automatic Coupler device for railroad cars. Mr. Jackson Beard was a farmer, businessman, railroad worker, blacksmith and inventor. In 1849 Andrew Jackson Beard was born into slavery in Alabama.

He spent the first fifteen years of his life as an enslaved person until slavery was abolished in 1865. Shortly after the end of slavery, Mr. Beard married and started a farm. After running a farm for about five years, he left the farming business to run a flour mill. It was during this time that Andrew Jackson Beard began thinking of a design for a new farm plow.

In 1881 he patented his first plow and sold the idea for $4,000. Soon after in 1887 he invented and patented another plow and sold that idea for $5,000.

With the money he made from his inventions he began a business in real estate.

Real estate is the business of buying and selling property such as land, houses, buildings, etc. Beard also began working with and studying engines. In 1892 Mr. Jackson Beard patented an improvement for the rotary steam engine. Finally while working on the railroads, he came up with his most known contribution- The Jenny Automatic Car Coupler. During that time railroad cars would need to be attached by hand and this was very dangerous. A railroad worker would stand in between the train cars as a conductor backed up the train.

The worker would have to wait to the exact moment the coupler was in place and drop in a pin to secure the two cars together. Many times the hands, arms, and legs of workers would get caught in between the cars injuring the workers. Some would even die from being caught between trains. Mr. Beard himself lost a leg in one of these accidents.

In 1897 he invented an Automatic Car Coupler also known as the Jenny Coupler. He did not invent the automatic coupler but his invention was an improvement on the automatic coupler invented by Eli H. Janney in 1873. Mr. Beard's car coupler made attaching railroad cars together much safer. It prevented injuries and saved many lives. Andrew Jackson Beard died in 1921. He was inducted into the National Inventors Hall of Fame in 2006.

Chapter 3
COLLIDING OBJECTS
Getting to the Core

ANDREW JACKSON BEARD
Mr. Beard's invention improved the way railroad cars were joined making it safer. Andrew Jackson Beard knew that joining train cars involved a small collision at very low speeds. How did Beard use the science of colliding objects in his invention?

CORE IDEAS
• When objects collide, energy transfer changes the objects' motion.

• Energy can be transferred in various ways and between objects

ENERGY AND COLLISION
Every day we experience and use the science of colliding objects in many parts of our life. Did you knock on the door this morning to tell your brother or sister to hurry up? Have you ever bumped the car door with your body to close it because your hands were full? Do you play sports such as baseball, softball, soccer, tennis, volleyball, cricket, football? All of these activities involve colliding objects.

MOMENTUM
Whenever an object is moving it has momentum. We also now know that the more speed an object has when it is moving the more energy it has. So if you throw a beach ball and a baseball with the same amount of force, will they both have the same amount of energy? The answer is no! A beach ball will not have the same amount of energy because no matter how hard you throw it, it will not travel as fast as a baseball, and speed = energy. This is because of the mass of each object is different. Mass is how much matter an object has, usually determined by weight. The more weight the more mass an object has.

In our example the beach ball is mostly air and plastic while a baseball is a solid object that weighs more. So when you throw both balls with the same force the baseball has more momentum. Momentum is a way to measure and think about an object's ability to stay in motion. Or for the object to keep moving at a certain speed.

Learning Exercise

MOMENTUM EXERCISE
Last week Marcus went grocery shopping with his mother. As they were loading the groceries into the car their shopping cart rolled away. Before Marcus or his mother could notice the cart picked up speed and collided with another car in the parking lot. Circle the cart that would have caused the most damage to the other car and explain the reason for this.

_____
_____
_____
_____
_____
_____
_____
_____

Chapter 3
COLLIDING OBJECTS
Getting to the Core

Let us take another example to really get the hang of mass and momentum. If someone had a giant bag of leaves and a huge bolder, and rolled them both down a steep hill, which would have more momentum? The bolder would have more momentum because it has more mass or more weight. This means the bolder would have more speed than the leaves and would be harder to stop from rolling because the bolder now has more kinetic energy.

COLLISION

Once an object is in motion it has momentum, which is a measurement of the ability of the object to stay in motion. What happens if our moving object hits another object? This is called a collision. Collisions may occur between a moving object and another object that is standing still. A collision can also happen between two moving objects. We learned in the last chapter that energy can be transferred between objects by motion. Whenever there is a collision, energy is transferred from one object to another. This happens in the brief moment when the objects touch each other during the collision. When a collision happens the transfer of energy causes changes in the momentum and direction of the objects involved. The energy that the objects have before the collision will be the same energy after the collision. The change will be in where the energy is located. The momentum may be transferred to another object or in another direction but the total momentum must remain the same. For example when a basketball player is dribbling the ball there is a collision between the ball and the floor. The ball returns to the players hand because the momentum is transferred between the ball and the floor sending the ball back in the opposite direction but with equal momentum.

Ball colliding with racquet

IMAGE	COLLISION	ENERGY TRANSFER RESULT
	The moving tennis racquet collides with the moving tennis ball.	The energy transfer from the collision results in sound and motionas the tennis ball is sent in theopposite direction.
	The motion of swinging the hammer creates momentum. The hammer collides with the head of the nail.	When the hammer collides with the nail the kinetic energy is transferred to the nail pushing it into the wood.

Chapter 3
COLLIDING OBJECTS
Writing In Science

Complete the chart below by drawing a picture explaining the collision and the result

PICTURE	COLLISIONS	ENERGY TRANSFER RESULT

Learning Exercise

We all use cars to get around from place to place. Unfortunately cars can be very dangerous when there is a collision. To keep drivers safe, roads have speed limits so people will not drive too fast. Write an essay explaining why speed limits help people stay safe if there is a car collision. Use the science of energy, speed, mass and momentum to explain how speed affects cars in a collision. Additional research may be done to complete the answer.

_____
_____
_____
_____
_____
_____
_____
_____
_____
_____
_____
_____
_____
_____
_____
_____
_____
_____
_____
_____
_____

Did You Know?
Self Driving Cars: The technology company Google has launched a new company called Waymo. Waymo is in the process of producing cars that drive themselves. These cars use computers, cameras, and sensors to control the car without a driver. Most automobile accidents are due to human error. Google is one of many companies hoping to have less collisions on the road by removing the driver. In Phoenix, Arizona people can take free rides around town—without a driver!!!

Chapter 3
COLLIDING OBJECTS
Science and the Invention

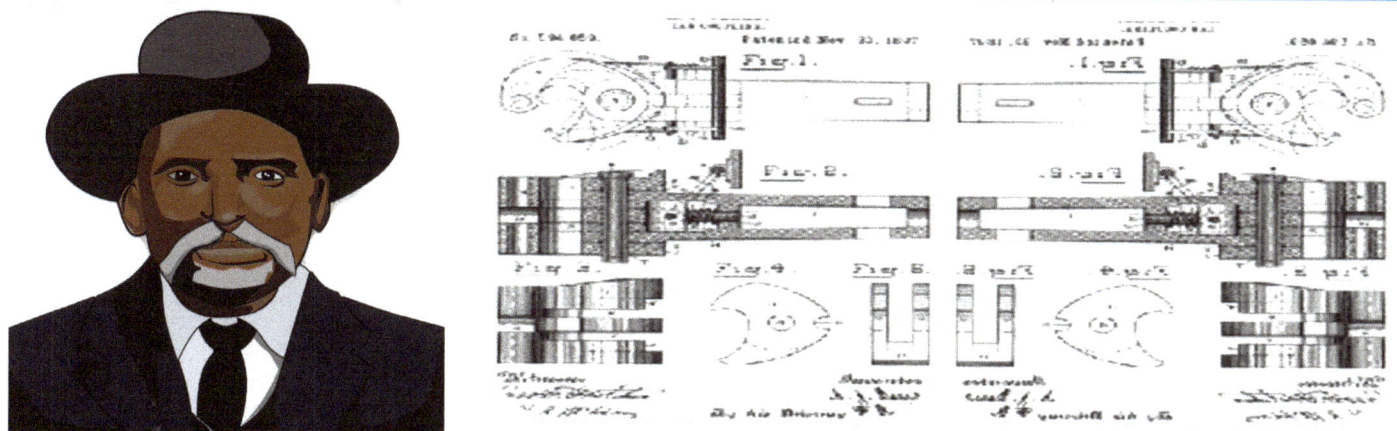

Our "Core Idea" tells us that when objects collide energy transfer changes the objects motion

- Attaching train cars together was a very dangerous job when it had to be performed by hand.

- Mr. Jackson Beard knew that when the train cars came together there was a small collision. Many railroad workers lost their arms, legs and even lives in these small collisions between the train cars.

- Beard used the energy of the trains colliding as they came together to transfer energy to the coupling device.

- Even though the trains collide at a low speed the mass of the trains gives the small collision a lot of momentum. This momentum is used to activate the automatic coupling device which links the train cars together safely

Investigation and Exploration Using the Principles of: Colliding Objects

Materials:
Various small and large marbles, 2 meter sticks, flat surface

Preparation:
- Lay the two meter sticks side by side leaving a space/track for the marbles in the middle.

Procedure:
- 1. Set the target marble on a marked point between the meter sticks
- 2. Shoot another marble at the target marble
- 3. Record how far the target marble moved from the point
- 4. Repeat with different combinations of target marbles and shooter marbles. Change the force used to shoot the marble at the target. Record the data.

PICTURE	DISTANCE
Small to Small	
Target Marble	
Big to small	
Small to small	
Big to Big	

Chapter 3
COLLIDING OBJECTS
Review Chapter

1. When objects collide:
a. the speed of the objects change
b. energy is transferred between objects
c. objects can change directions
d. all of the above

2. How does the mass of an object affect its momentum?
a. less mass increases an object's momentum
b. more mass decreases an object's momentum
c. the mass of an object does not affect the momentum
d. more mass increases an object's momentum

3. An object's ability to stay in motion is called
a. velocity
b. momentum
c. speed
d. kinetic energy

4. A heavy truck has more momentum than a car moving at the same speed because the truck:
a. is taller
b. has more mass
c. has larger wheels

5. When two or more moving objects transfer energy to each other for a short period of time it is called:
a. collision
b. work
c. momentum
d. energy

6. What is true about collisions?
a. the total momentum of the objects must remain the same
b. some momentum will be lost
c. the total momentum must remain the same
d. all of the above

7. Andrew Jackson Beard became good at many things in his life. Provide evidence from the text to explain why this statement is true.

_____
_____
_____
_____
_____
_____
_____
_____
_____
_____

8. Explain how Andrew Jackson Beard used the science of momentum and collisions in his Automatic Coupling device.

_____
_____
_____
_____
_____
_____
_____

Chapter 4

ENERGY CONSERVATION
How is energy converted from one form to another?

Chapter Vocabulary
- Conserve
- Energy conservation
- Energy transfer

Section Theme
Apply scientific ideas to design, test, and refine a device that converts energy from one form to another. 4-PS3-4

Objectives
- Explain the law of conservation of energy, from one form to another
- Explain the importance of the scientist's invention/innovation
- Demonstrate an understanding of the scientific concepts behind the invention

WILLIAM KAMKWAMBA

William was born in 1987 in the African country of Malawi. He grew up on a farm with his parents and six sisters. William completed his education up to the 8th grade.

When William was younger his country of Malawi experienced a famine or food shortage. William's parents were farmers and shortly after beginning secondary school William had to leave school because his parents did not have the money to pay his school fees due to a loss in income from the famine..

While William was not in school he began reading books at a local library. One of the books William read was an 8th grade text book about using energy. It was this book that gave William the idea to build the windmill. He continued reading and studying books about electricity and energy.

At the age of 14, using materials from a junkyard including a tractor fan, a broken bicycle, and blue gum trees, William built his first windmill. William was able to bring electricity to his home which powered four light bulbs and a radio. He was even able to charge his neighbors mobile phones.

William built several more windmills each better than the last. In order to help his family and his village with the famine he built a windmill which pumped water to the farms allowing the farmers to farm once again. William eventually finished his secondary education, and in 2010 he enrolled in Dartmouth College in the United States. William has also written an auto-biography entitled "The Boy Who Harnessed the Wind".

Chapter 4

ENERGY CONSERVATION
Getting to the Core

William Kamkwamba
was able to create a windmill that brought electricity to his village. How did William Kamkwamba use the science of energy conservation to bring electricity to his home and village?

CORE IDEAS
- The energy of motion can be converted into electrical energy.
- Energy can be transferred in various ways between objects.

CONSERVATION OF ENERGY

You may have heard of "energy conservation". This is when are saving electricity, using less water, recycling and other things that lower the amount of energy we use. The law of conservation of energy is one of the laws of science. It is very different from using less energy around the house.

We learned that energy can be transferred through light, sound, heat and electricity. Energy transfer takes place because energy is conserved. Conserve means to keep the amount or the quantity the same. The law of conservation of energy says that energy is not created or destroyed, but only changed from one form to another. This means that even though the type of energy might change the amount of energy transferred stays the same.

Let's go through an example to help us understand better. Cars run on gasoline. The gasoline's chemical energy is released when it is burned. The chemical energy changes into thermal energy or heat. The thermal energy is converted to mechanical energy causing the engine parts to move. The moving parts in the engine connect to the cars wheels and make it go.

burned gasoline changes to thermal energy to power a car

When a driver steps on the gas pedal more gas is sent into the engine or more chemical energy. This means more heat energy is created in the engine. The increase in heat energy is transferred and creates more mechanical energy. As the mechanical energy in the engine is increased the speed of the car increases. In this example we can see how the energy is changing from one form to another because it is being conserved each time it changes.

 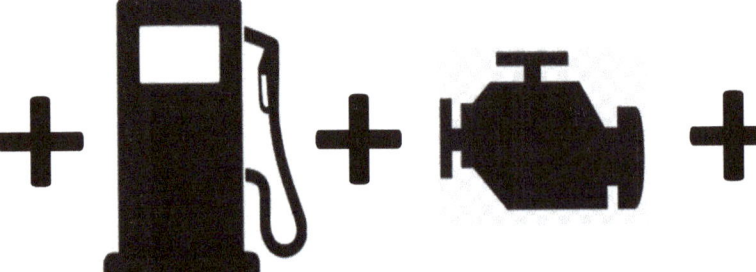

Chapter 4: ENERGY CONSERVATION
Writing In Science

Write a short story. Your story, for example can be about home, school, a camping trip, science fiction or fantasy, etc.. Give at least 5 examples of the law of conservation of energy and energy transfer in your story. Be creative!

Did You Know?
Ocean Wave Energy: We are learning about devices that convert energy from one form to another. A new device that is being tested is a huge buoy (boo-ee) that turns ocean waves into electricity. A buoy is a floating device that can have many uses such as helping boats know where they are. This device is called CETO. The CETO uses the up and down motion of the ocean waves to move parts in the buoy which then creates electricity.

Chapter 4
ENERGY CONSERVATION
The Science & the Invention

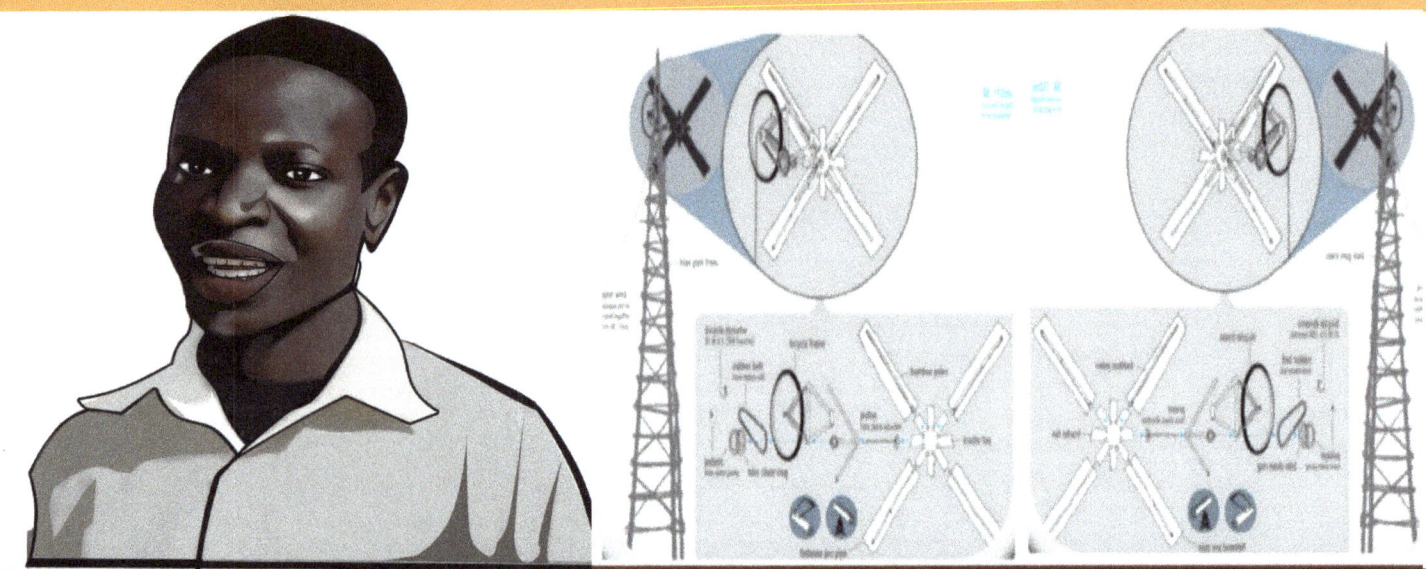

Our core idea tells us that: Energy can be transferred from place to place by electric currents. The currents may have been produced by transforming the energy of motion into electrical energy.

- William Kamkwamba used what he knew about the conservation of energy and energy transfer to build his windmill.

- William knew that the kinetic energy of the wind would make the blades of the windmill spin creating mechanical energy.

- William used the energy of the spinning blades that is conserved and transferred it into an electrical current.

- The electric current was then used to power homes in his village.

- William Kamkwamba was not the first person to build a windmill. What makes William's windmill special is that he used creativity. Using materials from a junkyard he was able to transfer wind energy into electricity and help his family and village.

Investigation and Exploration Using the Principles of: Conservation Energy

Experiment: Lemon Battery
Materials:
4 Lemons, 4 galvanized nails, 4 pieces of copper (pennies),
5 alligator clips, a small LED light to power up

Procedure:
1. Roll the lemon to release the juice inside

2. An adult should assist with cutting a small slit into each lemon for the penny.

3. Insert one nail and one piece of copper into each lemon.

4. Connect the ends of one wire to a galvanized nail in one lemon and then to a piece of copper in another lemon. Do this with each of your four lemons until you have them all connected. When you are finish

5. Connect the unattached piece of copper (positive) and the unattached nail (negative) to the positive and negative connections of your light. The lemon will act as the battery.

QUESTIONS
- What caused the light to work? Was there energy being transferred or changed from one form to another?

- Name at least three energy types that are used or produced **in this experiment.**

Chapter 4: ENERGY CONSERVATION
Chapter Review

1. Which is an example of the law of conservation of energy?
a. Hitting a baseball
b. Turning off a light when you leave the room
c. Not running the water while brushing your teeth
d. Riding a bike instead of driving

2. The law of conservation of energy states:
a. every action has an equal and opposite reaction
b. energy is created and then destroyed
c. energy is not created, it is continuously changed from one form to another
d. energy cannot be transferred and must be saved

3. When the wind spins a windmill, the wind's energy is conserved when: (select all that apply)
a. the windmill's blades turn a turbine
b. electricity is transferred to power homes
c. electricity is produced by a turbine
d. only (a) and (c) are correct
e. all of the above

4. What does it mean to conserve?
a. to only use one form of energy
b. to not waste or overuse energy and resources
c. to use different forms of energy
d. to keep the amount or quantity the same
e. (b) and (d) are correct

5. Explain how William Kamkwamba's love of reading helped him develop his windmill invention.

_____
_____
_____
_____
_____
_____
_____
_____
_____
_____
_____
_____

6. he word ingenuity means being clever, original, and inventive. How did William Kamkwamba use ingenuity in the creation of his invention?

_____
_____
_____
_____
_____
_____
_____
_____
_____
_____
_____
_____

ANSWER KEY

Speed and Energy Chapter Review
1. b
2. c
3. d
4. a
5. b
6. a
7. Answers should mention the rocket he created and Johnson winning the national science competition
8. Answers may include the transfer of energy from the hand pump to the water chamber

Energy Transfer Chapter Review
1. d
2. d
3. a
4. d
5. a
6. c
7. Answers should mention the rocket he created and Johnson winning the national science competition

8. Answers may include the transfer of energy from the hand pump to the water chamber

Colliding Objects
1. d
2. b
3. b
4. a
5. a
6. Answer should include accomplishments as a farmer, businessman, railroad worker, and/or blacksmith
7. Answer should include explanation of energy transfer

Conservation of Energy
1. a
2. c
3. e
4. e
5. c
6. Answers should include Kamkwamba's studies at the library
7. Answers should include Kamkwamba's use of junk materials to build his invention

REFERENCES

CONTENT
- Biography.com Editors, (2014, April). Lonnie G. Johnson Biography Retrieved from http://www.biography.com/people/lonnie-g-johnson-17112946#early-life-and-education
- Chamberlain, G. (2012, November). Patricia Bath . Retrieved from http://blackinventor.com/patricia-bath/

CrashCourse (June 2, 2016). Collisions: Crash Course Physics #10. Retrieved from https://www.youtube.com/watch?v=Y-QOfc2X-qOk
- Fink, J. (2013, May). How to Make a Lemon Battery. http://kidsactivitiesblog.com/28028/lemon-battery
- Kamkwamba, W. The Boy Who Harnessed the Wind. Retrieved from https://williamkamkwamba.typepad.com/about.html
- Kirkwms (February 1, 2015). Andrew Jackson Beard 1 of 10. Retreived from https://www.youtube.com/watch?v=y_LMbncGxjc
- Holmes, K (2008). Black Inventors: Crafting Over 200 Years of Success. Brooklyn, NY. Global Black Inventors Research Project

myblackhistory.net. Andrew Jackson Beard. Retrieved from http://www.myblackhistory.net/Andrew_Jackson_Beard.htm

Thompson, V (2018, March). Cataract Surgery. Retrieved from http://www.allaboutvision.com/conditions/cataract-surgery.htm
- The HistoryMakers. (2012, November). Dr. Patricia Bath. Retrieved from http://www.thehistorymakers.com/biography/dr-patricia-bath
- Victor and Kellough, (2004) Science K-8 An Integrated Approach

(2017, February). Shanghai Maglev Train — The Fastest Train in the World. Retrieved from http://www.chinahighlights.com/shanghai/transportation/maglev-train.htm

PHOTO CREDITS
- Shanghai high speed Maglev Train Image by Somkiat Atthajanyakul | Dreamstime.comShanghai high speed Maglev Train Image by Somkiat Atthajanyakul | Dreamstime.com - Page 9

- Smart Car: https://www.thenational.ae/business/autonomous-travel-smart-cars-dumb-drivers-1.96576 - Pg. 20

- Electromagnetic Spectrum Diagram By Designua |Shutterstock - Page 13

- Thermal Energy Diagram - Source unknown/no-longer available online - Pg 13

- Ocean Wave Eneregy Source - Pg 25

By Carnegie Wave Energy Limited - Supplied by Carnegie Wave Energy Limited., CC BY-SA 3.0, https://commons.wikimedia.org/w/index.php?curid=31753495

- Bio-Battery ID 92503877 © Haryigit | Dreamstime.com - Pg. 26

- Lonnie Johnson - Pg.28
https://www.bbc.com/news/magazine-37062579

- Dr. Patricia Bath - Pg. 28
https://www.ossc.org/content.aspx?page_id=22&club_id=239344&module_id=79755

- Andrew Jackson Beard - Page 28
https://www.pinterest.ca/pin/169448004700670281/

- William Kamwamba - page 28
https://www.amazon.com/William-Kamkwamba/e/B002BRKQEK%3Fref=dbs_a_mng_rwt_scns_share